# EUREK-HER!

## STORIES of INSPIRATIONAL WOMEN in STEM

b small

For all the brilliant women in my life. - **F.D**

For my son Aris, my wife Andrea, my mum Rosa and my uncle Cinto. - **N.V**

**b small**

Published by b small publishing ltd.
www.bsmall.co.uk

Text and illustrations copyright © b small publishing ltd. 2024

1 2 3 4 5
ISBN 978-1-913918-94-1

Publisher: Sam Hutchinson
Art director: Vicky Barker
Editorial: Alice Harman

Printed in China on paper from responsible sources.

All rights reserved.

No reproduction, copy or transmission of this publication may be made without written permission. No part of this publication may be reproduced, stored in a retrieval system or transmitted in any form or by any means, electronic, mechanical, photocopying, recording or otherwise, without the prior permission of the publisher.

British Library Cataloguing-in-Publication Data.
A catalogue record for this book is available from the British Library.

# CONTENTS

**6–7
TAPPUTI
BELATEKALLIM**
(MESOPOTAMIA, c. 1200 BCE)

**8–9
MARY
HEBRAEA**
(ALEXANDRIA, c. 1ST CENTURY CE)

**10–11
MARIA
SIBYLLA MERIAN**
(GERMANY, 1647–1717)

**12–13
SARAH
GUPPY**
(ENGLAND, 1770–1852)

**22–23
MADAM C.J.
WALKER**
(USA, 1867–1919)

**24–25
LISE
MEITNER**
(AUSTRIA, 1878–1968)

**26–27
INGE
LEHMANN**
(DENMARK, 1888–1993)

**28–29
MARGUERITE
PEREY**
(FRANCE, 1909–1975)

**38–39
TU YOUYOU**
(CHINA, BORN 1930)

**40–41
ROSELI
OCAMPO-FRIEDMANN**
(PHILIPPINES, 1937–2005)

**42–43
OMOWUNMI
SADIK**
(NIGERIA, BORN 1964)

**44–45
NZAMBI
MATEE**
(KENYA, BORN 1992)

**14–15**
**JEANNE VILLEPREUX-POWER**
(FRANCE, 1794–1871)

**16–17**
**MARY ANNING**
(ENGLAND, 1799–1847)

**18–19**
**MARGARET KNIGHT**
(USA, 1838–1914)

**20–21**
**MARGARETE STEIFF**
(GERMANY, 1847–1909)

**30–31**
**BIBHA CHOWDHURI**
(INDIA, 1913–1991)

**32–33**
**ROSALIND FRANKLIN**
(ENGLAND, 1920–1958)

**34–35**
**STEPHANIE KWOLEK**
(USA, 1923–2014)

**36–37**
**VERA RUBIN**
(USA, 1928–2016)

## 46–65 ACTIVITIES TO DO AT HOME

| | | | |
|---|---|---|---|
| 46–47 | DRAW A REALISTIC FLOWER | 56–57 | MAKE A POND VIEWER |
| 48–49 | MAKE YOUR OWN FOSSIL | 58 | PLAY THE 'SMELL TEST' GAME |
| 50–51 | DESIGN A BRIDGE | 59 | BUILD A RECYCLED TOY HOUSE |
| 52 | MAKE PERFUME – FOR YOUR HOME! | 60–61 | DESIGN YOUR OWN PERIODIC TABLE |
| 53 | SET OFF A CHAIN REACTION | 62–63 | MEASURE SOMETHING YOU CAN'T SEE |
| 54–55 | MAKE A PAPER BAG | 64–65 | MAKE A LAVA LAMP |

**66–71 GLOSSARY**     **72 LIST OF PERIODIC ELEMENTS**

# TAPPUTI BELATEKALLIM
## (MESOPOTAMIA, c. 1200 BCE)

Tapputi lived in the Middle East, in a kingdom called Assyria, more than 3,000 years ago. Everything we know about her comes from a few lines of ancient writing carved into a broken stone **tablet**. But what we do know is pretty amazing!

### FIRST PERFUME MAKER

In the stone tablet, Tapputi is called 'muraqqītu' – a person who makes **perfume**. She is the earliest perfume maker to ever be named, in all of history! The tablet describes Tapputi's perfume as being 'fit for a king', and she must have been important to be named at all.

Perfumes were very important in the ancient world. Often, only very rich people could afford to buy them. Tapputi would have thought of herself as a perfume maker, not a scientist. But the stone tablet describes Tapputi using a scientific process called **distillation** for the very first time recorded in history!

*We know that Tapputi was a woman because her name and title, 'muraqqītu', are both written in the feminine form of those words.*

Distillation separates out parts of a mixture by heating liquid until it turns to **vapour** and then cooling it back into liquid. In perfume making, this process captures the nice-smelling essential oils from flowers and other ingredients. People often call Tapputi the world's first known **chemist** because chemists often use distillation.

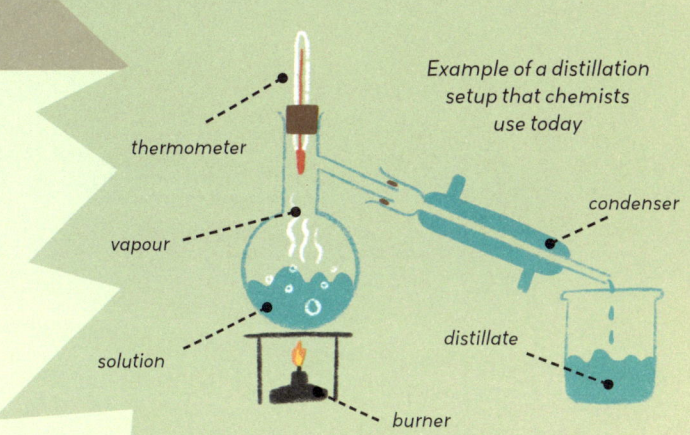

*Example of a distillation setup that chemists use today* — thermometer, vapour, solution, burner, condenser, distillate

## CHEMIST

A chemist studies chemistry, the science of what things are made of and how they work together.

## THE SCIENCE OF SMELLS

Tapputi's stone tablet includes her recipe for turning flowers, oils, tree resin, water and other ingredients into a perfumed lotion. First, she boiled her chosen ingredients together in a device called a still. She then left the mixture overnight.

In the morning, she **filtered** the mixture to remove unwanted material. Then she began the distillation process again, and kept repeating it until she had made the perfect perfume!

Today, Tapputi still inspires many perfume makers who use natural ingredients rather than artificial chemicals.

*Turn to page 52 to make perfume for your home!*

*The stone tablet tells us that Tapputi worked with another woman, who we know only as '-ninu'. The first part of her name has chipped off.*

# MARY HEBRAEA
## (ALEXANDRIA, c. 1ST CENTURY CE)

Mary lived in the ancient Egyptian city of Alexandria around 2,000 years ago. She founded a school of alchemy and experimented with chemical reactions.

Alchemy is the science of trying to find ways to live for ever and to turn cheap metals into gold. We now know that both these things are impossible, but many alchemists such as Mary gained important scientific knowledge from their work.

### DISTILLATION

Mary invented more advanced equipment to use for **distillation** and for boiling liquids at high temperatures. She seems to have also come up with the idea of using a paste of flour and water to tightly seal the containers she used for experiments. This simple, effective solution was used for centuries.

Mary's most famous invention is a double boiler. This heats water in an outer container to gently warm a substance in the inner container and keep it at the same temperature. The double boiler is still used in chemistry and cooking today.

A double boiler is often known as a 'bain-marie', which means 'Mary's bath' in French.

## AN UNUSUAL ALCHEMIST

Alchemy is sometimes described as an early form of what eventually became **chemistry**. Mary created experiments to help her understand how different materials worked and how they behaved when you put them together, rather than always focusing on turning them into gold.

Some historians believe that Mary invented a process to make a **compound** called silver sulphide. Metalworkers still use this today to help them engrave (scratch) designs into metal.

### COMPOUND

A compound is a substance made of two or more different **elements**. Elements – such as iron and oxygen – are the basic materials that make up everything in the universe (except energy).

A lot about Mary's life will probably remain a mystery. But the processes and equipment that she likely invented are still helping scientists, chefs and others two thousand years later.

*Mary is also known as Maria and Miriam. She is sometimes called 'Mary the Jewess', which is an old-fashioned term for a Jewish woman.*

# MARIA SIBYLLA MERIAN
## (GERMANY, 1647–1717)

Maria's stepfather was very famous for painting flowers and insects, and encouraged Maria to become an artist.

Maria also grew up with a love of nature. She collected insects to study them and see how they changed through their **life cycles**. She watched as caterpillars transformed into different butterflies and moths, and painted the amazing things that she saw.

### NEW KNOWLEDGE

When she grew up, Maria illustrated two popular books full of pictures of flowers. Her third book included prints of the **metamorphosis** of caterpillars from eggs into butterflies.

At Maria's time, people didn't really understand insects' life cycles. They often believed that they suddenly appeared from mud, sand or slime. Maria proved that insects change at different stages of their lives.

Maria was the first person known to combine painting with the study of living things. Her artworks helped people understand her work as an entomologist (insect expert).

# NEW ADVENTURES

When Maria was 52, she and her daughter joined an **expedition** to Suriname, in South America. At this time, it was very unusual for women to travel so far to work as scientists. They spent two years painting Suriname's flowers and wildlife.

Maria and her daughter brought home pictures of plants, spiders, butterflies and lizards that people in Europe had never seen before.

At least six plants, nine butterflies, two bugs, one spider and one lizard have been named after Maria.

Maria passed on her knowledge to her two daughters, who also created wonderful scientific illustrations.

Can you investigate and draw something in nature? Turn to pages 46–47.

# SARAH GUPPY
(ENGLAND, 1770–1852)

Sarah was born into a very rich family and was given an excellent education. She married a man who owned a company that made machinery. This was at the time of the **Industrial Revolution**, when new machines and technologies were changing how people lived.

Sarah got involved in her husband's business, teaching herself about the different ways that machines could work. She talked about her ideas with other brilliant **engineers** and **inventors**, and came up with inventions of her own.

## BUILDING BRIDGES

Sarah designed a bridge that would not be washed away in a flood. It used metal chains to help hold it up. Sarah's design was never built but she **patented** this invention, and others, which meant that the government officially recognised them as her designs.

A few years later, Sarah helped to design a bridge in Wales. She chose not to be paid for this work, preferring the money to be spent on building a bridge that everybody could use. She also invested in her friend Isambard Kingdom Brunel's design for the famous (and still standing) Clifton Suspension Bridge in Bristol, where she lived.

## INGENIOUS IDEAS

Through her life, Sarah registered ten patents for her inventions. These included: a bed that doubled up as an exercise machine, a tea maker that could also cook eggs and keep toast warm, and a candlestick that helped candles burn brighter for longer.

*Sarah's idea for a bed that doubled up as an exercise machine*

One of Sarah's best ideas was to plant trees next to train tracks to prevent landslides. This is still done today! She also found a way to stop sea creatures called barnacles attaching themselves to ships, and sold it to the Royal Navy for £40,000.

Alongside her inventing, Sarah campaigned and raised money for lots of different charities, to help other people to have better lives. She even wrote children's books and then donated the profits to her favourite charities.

*Can you engineer your own bridge? Turn to pages 50–51 for an experiment inspired by Sarah's inventions.*

# JEANNE VILLEPREUX-POWER
## (FRANCE, 1794–1871)

Jeanne grew up in a tiny village in France and had a very basic education. She became a dressmaker, and was so successful that she was asked to make a princess's wedding dress.

When Jeanne got married, she and her husband moved to Sicily – an island off the south coast of Italy. They lived in a city beside the sea and Jeanne loved exploring her new surroundings. She was fascinated by the amazing nature of Sicily and wanted to understand it better.

### BECOMING A SCIENTIST

Jeanne began to teach herself about the science and history of nature. She read books, wrote letters to top scientists, and collected butterflies, rocks and shells.

Jeanne asked local fishermen to bring her interesting sea animals to study, but it was very difficult for her to see them under the water. She didn't want to take the animals out of the water to study them, as they would soon die. What could she do?

## INVENTING AQUARIUMS

Jeanne solved her problem by building a glass box that she could fill with water. She put the sea creatures inside to study them up close. Her invention was later called an aquarium.

Jeanne invented three different types of aquarium – one to use on land, and two to attach to the ocean floor.

## SECRETS AND SHIPWRECKS

Jeanne learned amazing new things about sea creatures. She watched octopuses using stones as tools to find food, and she was the first person to see that the paper nautilus octopus grew its own shell.

Most of Jeanne's research work was tragically lost in a shipwreck. Luckily, she had already shared some of her knowledge in books and letters. Scientists – and pet fish owners! – all over the world still use Jeanne's invention today.

*The scientific study of sea creatures and plants, and their environments, is called marine biology.*

*Can you explore our underwater world? Turn to pages 56–57 for an activity inspired by Jeanne.*

15

# MARY ANNING
(ENGLAND, 1799–1847)

Mary grew up close to the sea, in a place called Lyme Regis. Mary's father taught her how to find and clean the **fossils** buried in the rocky cliffs and beaches. The family was very poor, so they all hunted for fossils to sell to tourists from a table outside their home.

### FOSSILS
Fossils are the remains or traces of living things from over 10,000 years ago. They are often formed from hard parts of animals, such as bones and teeth, which have turned into rock.

When Mary was just 12 years old, she **excavated** a complete, 5.2-metre-long skeleton of an *Ichthyosaurus* – a reptile that swam in the sea at the time of dinosaurs. Mary spent months carefully digging it out of the rocks.

*When Mary was a baby, she survived being struck by lightning!*

## SELF-TAUGHT SCIENTIST

Mary became an expert in fossils, and kept careful records of everything she found. She taught herself the sciences of **geology** and **anatomy** to better understand her discoveries.

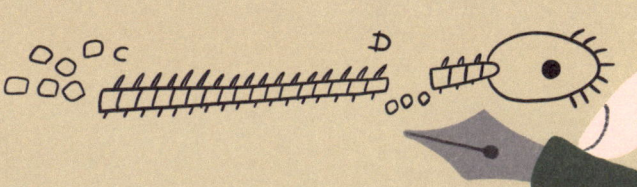

When Mary made her most famous discovery, the first complete skeleton of a *Plesiosaurus*, some scientists thought it must be a fake! They eventually admitted their mistake.

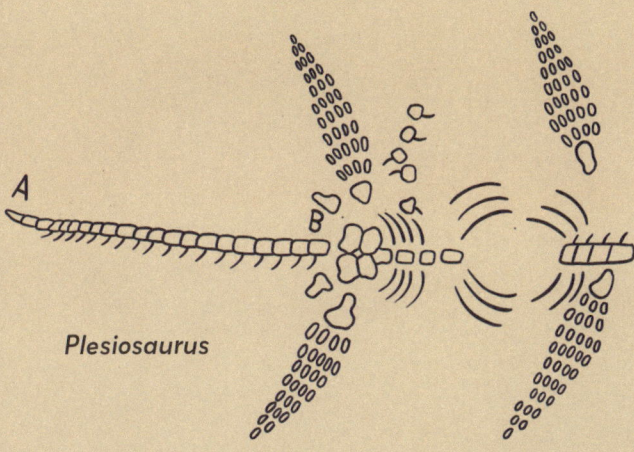

*Plesiosaurus*

Many scientists came to see Mary's work, and she often knew more about palaeontology (the study of fossils) than them. However, because Mary was from a poor family, they didn't mention her in their research. Not even when she'd discovered and identified the fossils they were writing about!

## UNDERSTANDING THE PAST

The fossils Mary found often ended up on display in museums. Huge crowds came to see them and learn about the past. Mary's work also helped other scientists understand that animal **species** can go extinct (die out forever).

In her lifetime, Mary never got the respect that she deserved for her great work, and she was always poor. However, since Mary's death, people have started to give her proper credit for her amazing discoveries.

*Mary was also a leading expert in the study of coprolites ... that's fossilised poo!*

*Turn to pages 48–49 to learn how to make your own fossils.*

# MARGARET KNIGHT
(USA, 1838–1914)

As a child, Margaret loved to play with tools and make things. Her family was very poor, so she left school at 12 years old to work in a **cotton mill**.

When someone was hurt in an accident using equipment at the mill, Margaret invented a safety device to stop it happening again. But she didn't know that she could **patent** her design. This meant she didn't get paid for it, even when many other mills started using it.

### PATENT
When you apply to patent your invention, you're asking the government to recognise you as its **inventor**. You can then stop others from making, using or selling your invention for a certain number of years.

*Margaret was known as Mattie.*

## PAPER BAGS AND PATENTS

Margaret got a job at a **factory**, folding paper bags by hand. She knew there must be a quicker, better way so she invented a machine that automatically cut and folded flat-bottom paper bags. This type of paper bag was stronger and easier to carry, and difficult to make by hand.

Margaret applied for a patent for her machine, but a man tried to steal her idea and apply for the patent himself! He claimed no woman could design such a machine, but Margaret presented detailed notes and hand-drawn diagrams to show she had. She won!

## A LIFETIME OF INVENTIONS

Margaret formed her own paper bag **company**, and continued inventing for the rest of her life. She registered at least 25 more patents – including designs for a skirt protector, a window frame and a machine for making shoes. She never made a huge amount of money, but she wasn't poor.

In 2006, Margaret was awarded a place in the Inventors' Hall of Fame. Flat-bottom paper bags are still very popular around the world today.

*In 2020, fewer than 1 in 6 international patent applications were made by women. Still!*

*Turn to pages 54–55 to learn how to make your own paper bag.*

19

# MARGARETE STEIFF
(GERMANY, 1847–1909)

When Margarete was young, she was very ill with a disease called **polio**. It meant she could not walk and her right arm always hurt. Her sisters took her to school in a cart every day, and she worked hard to learn how to sew very well.

She and her sisters saved up enough money to buy a sewing machine – the first in their town. When her sisters got married, she carried on their business alone, and eventually opened her own clothes shop.

## INVENTING SOFT TOYS

One day, Margarete made a little **felt** elephant filled with soft wool. She liked it, and made more as gifts for the children in her family.

The children loved playing with them. Margarete had invented the first known soft toy! She started selling the elephants, and other animals including rabbits dogs, cats and mice.

The toys were so successful that Margarete set up a **factory** and employed people to help make them. Her nephew designed a toy bear with arms and legs that could move. These bears became known as 'teddy bears' and were hugely popular.

## GOING GLOBAL

Margarete was a smart businesswoman who overcame huge challenges, particularly as a disabled woman, to build a famous global **company**. In less than 15 years, her business grew from 14 employees to 2,200 people making over 2 million toys a year!

Margarete liked to say '*only the best is good enough for children*'. Today, her invention still brings happiness and comfort to children around the world. Do you have a favourite soft toy?

*Teddy bears are named after the US president Theodore ('Teddy') Roosevelt, because of a popular story about him refusing to shoot a bear.*

*A company in the USA created their own teddy bears around the same time as Steiff, but neither company knew about the other.*

*Steiff toys are still made at Margarete's factory in Germany today. They always have a metal button in their ear.*

# MADAM C. J. WALKER
(USA, 1867–1919)

When Madam C. J. Walker was born, she was named Sarah Breedlove. Her parents were African American people who had been **enslaved**. She was the first person in her family born after slavery was made illegal.

Sarah's family was very poor and she didn't go to school. However, she did learn to read a little at **Sunday school**. Her parents died when she was only seven years old, so she went to live with her older sister. They worked together in the cotton fields.

## FINDING HER PATH

By the time Sarah was 20 years old, she was a **widow** with a daughter. She worked very hard, cooking and washing clothes, to pay for her daughter to go to school.

Over time, Sarah became ill and her hair started to fall out. She used a hair product made by an African American businesswoman named Annie Malone. She liked it so much that she got a job selling it to other women. Sarah also married again and changed her name to Madam C. J. Walker.

# Mme. C. J. Walker

## STAR PRODUCT

Madam C. J. Walker invented her 'Wonderful Hair Grower' to heal **scalps** and encourage hair growth. The ingredients included sulphur, which is believed to strengthen hair, and moisturisers such as beeswax. A **perfume** made from violet flowers covered the sulphur's egg-like smell!

Madam C. J. Walker travelled all over the USA to sell her creation to African American women. It was so popular that she opened a **factory** and created a range of different hair products. She also opened a hair salon, a beauty school, and a **laboratory** for creating new products.

Madam C. J. Walker was the first female self-made millionaire in the USA. She used her wealth and influence to help others, supporting Black communities and the struggle for equal rights. She also gave lots of money to charity.

Madam C. J. Walker employed around 40,000 African American people to sell her products in the USA, Caribbean and Central America.

Madam C. J. Walker's **company** still exists today, and gives two-thirds of its profits to charity.

# LISE MEITNER
(AUSTRIA, 1878–1968)

When Lise was a child, girls in Austria could not stay in school after the age of 14. However, Lise had private lessons and worked hard to pass her exams. She became one of the first women to study a **PhD** in **physics** at the University of Vienna.

Lise then taught physics and studied **radioactivity** in Germany. She was part of a team that experimented with firing neutrons (tiny particles found inside **atoms**) at a chemical **element** called uranium. Lise had Jewish heritage so she had to leave Nazi Germany. She continued her research from her new home in Sweden.

### RADIOACTIVITY
Radioactivity is when an unstable substance releases energy in particles or waves.

### SPLITTING THE ATOM

Lise's experiments showed that atoms of uranium split apart when neutrons are fired at them. This releases a huge amount of energy. Lise explained how this process worked and she gave it the name nuclear fission.

Lise did not get proper credit for her work, and one of her coworkers won a **Nobel Prize** for the discovery. Many people believe that Lise should have shared this prize with him.

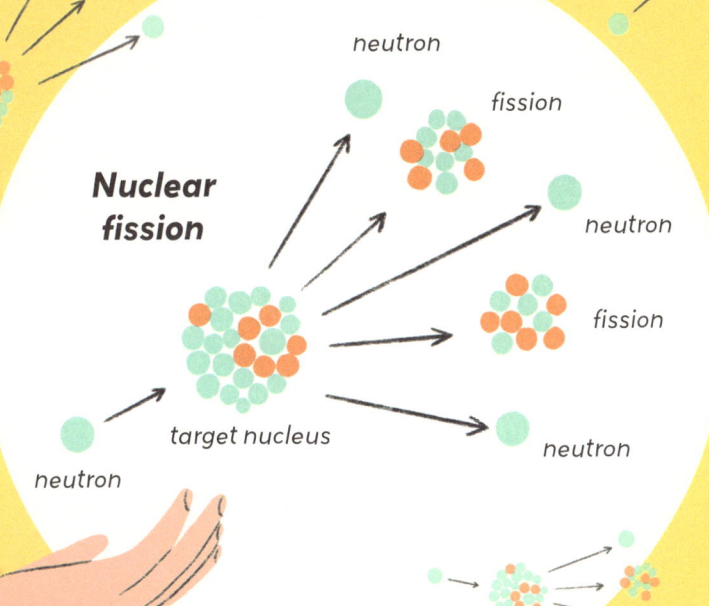

## NUCLEAR POWER

Lise realised that she'd discovered something incredibly powerful, and that this power could be dangerous. She was invited to work on creating deadly atom bombs, which used nuclear fission, but she refused. Two atom bombs were dropped on Japan during the Second World War, killing more than 200,000 people.

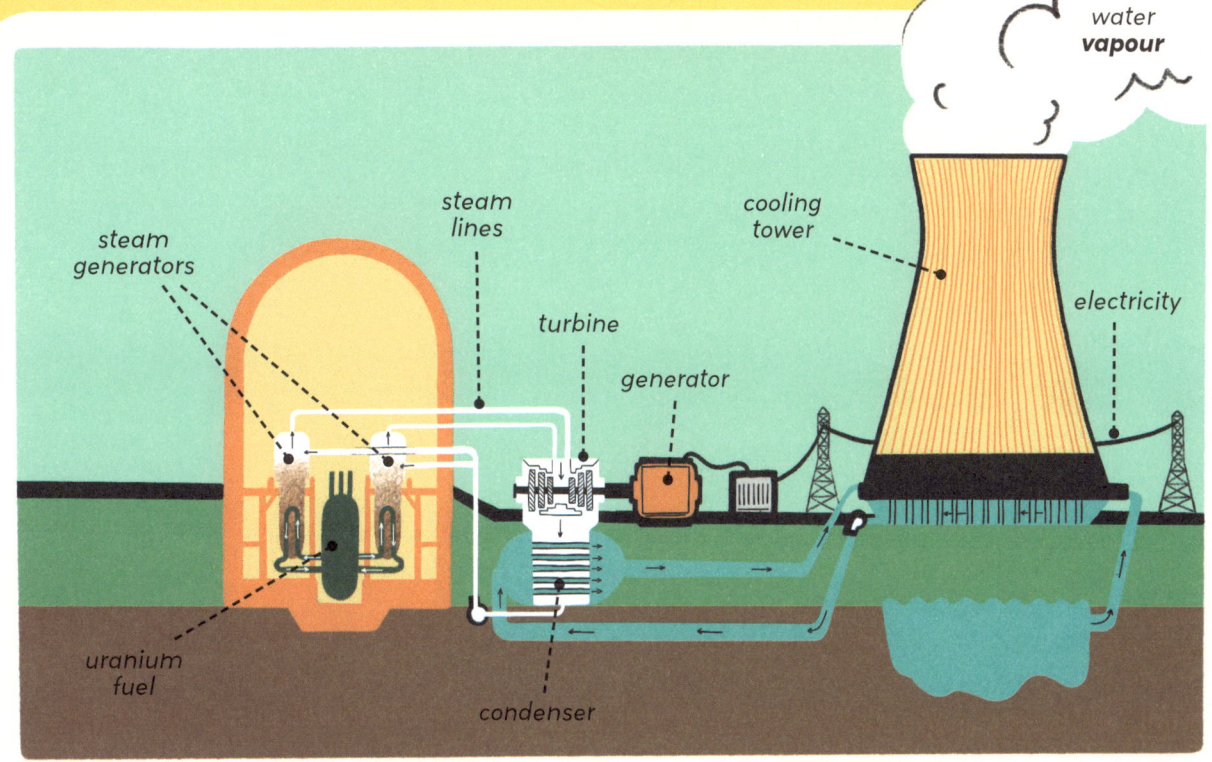

After Lise's discovery, people realised that energy from nuclear fission could be used to generate electricity. There are still concerns about the safety of this process, and the toxic waste it produces. However, many people believe it is the best realistic alternative to burning fossil fuels such as coal.

*Meitnerium, the 109th element of the **periodic table**, is named in Lise's honour.*

**109 Mt**

*Turn to page 53 to set off your own chain reaction – with dominoes!*

25

# INGE LEHMANN
## (DENMARK, 1888–1993)

Inge went to a school where boys and girls were treated equally and studied the same subjects. This was very unusual at the time. Inge was fascinated by mathematics and decided to study this at university.

Inge studied on and off for the next 15 years, taking time out due to illness and to work as an actuarial assistant. This job involved using her maths skills to help measure how likely certain things are to happen in future.

After Inge finished her studies, she eventually got a job where she learned about **seismology** and helped to set up the first stations in Denmark that measured seismic activity.

### SEISMOLOGY

Seismology is the study of **earthquakes** and seismic waves. A seismic wave is a vibration that travels through a planet. It is caused by events such as earthquakes and **volcanic eruptions**.

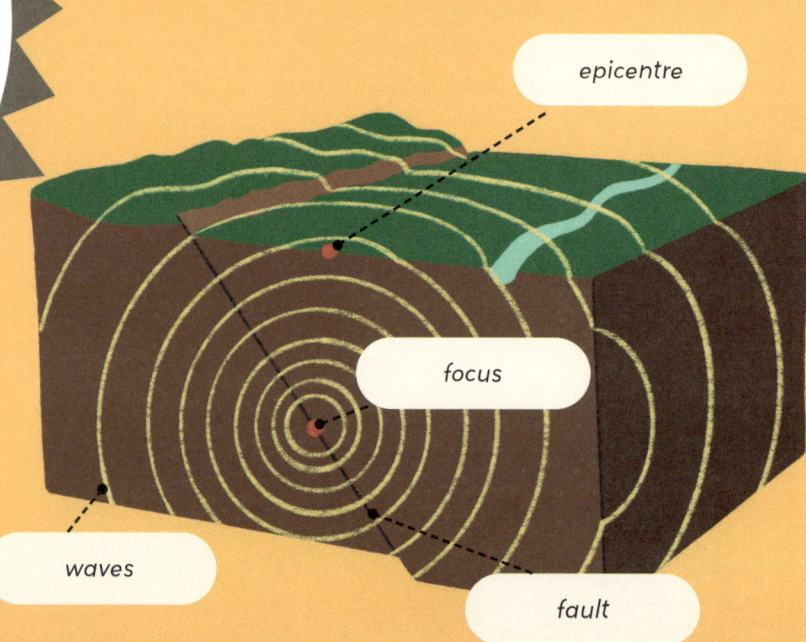

epicentre

focus

waves

fault

## SUPER SEISMOLOGIST

Inge loved seismology so much that she went back to university to study it. Very soon after she graduated, she was put in charge of an entire seismology organisation! She studied the effects of earthquakes, and how different seismic waves travelled.

One day, the waves from an earthquake in New Zealand did not behave as Inge expected. Some of the waves seemed to have been 'bent' and started travelling in different directions. Inge realised that the waves must have bounced off something solid deep inside the Earth.

## RETHINKING EARTH

Inge had discovered the inner core at the very centre of Earth. Over 40 years later, new technologies proved that her ideas were correct.

Inge's discovery changed the way that scientists understand the Earth's structure and how several of its systems work. Inge carried on researching through her life, and made one of her most important discoveries – a part of Earth's upper mantle where seismic waves travel more quickly – at the age of 76.

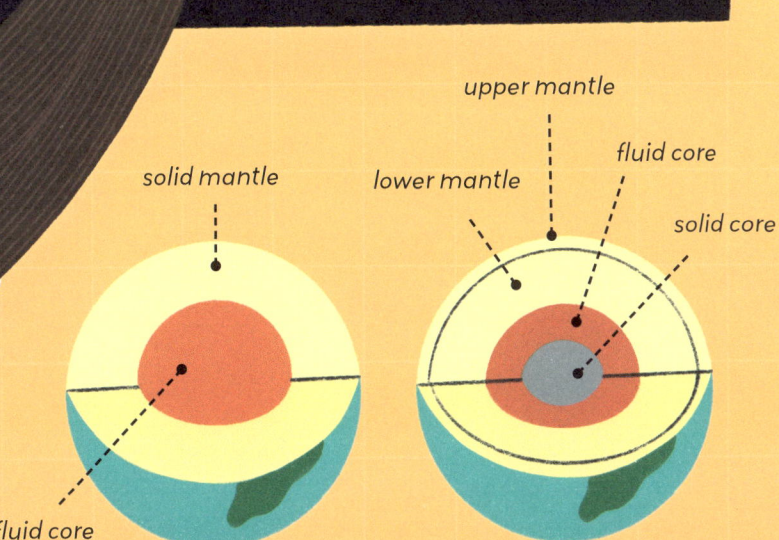

**Pre-Inge Earth**: solid mantle, fluid core

**Post-Inge Earth**: upper mantle, lower mantle, fluid core, solid core

*Turn to pages 64–65 to make an Inge Lehmann-inspired lava lamp!*

Earth's inner core is an extraordinarily hot, mostly metallic ball that's around 1,200 km wide. This is around the distance from London, UK to Barcelona, Spain.

27

# MARGUERITE PEREY
## (FRANCE, 1909–1975)

When she was young, Marguerite wanted to be a doctor. But her family was too poor for her to go to medical school. She became a **laboratory** assistant, working with the chemist Marie Curie.

Curie was a famous scientist, at a time when women in science rarely got the respect they deserved. Marguerite's job was to prepare **radioactive elements** for Marie's experiments.

### RADIOACTIVE ELEMENTS

An element – such as gold or oxygen – is a substance that can't be broken down into simpler substances. Some elements are radioactive, which means they always change over time, giving off energy.

### A RARE FIND

One day, Marguerite noticed a radioactive element behaving unexpectedly. Her tests proved that the element was changing into an unknown substance. Marguerite had discovered a new element! She called it francium, after her home country of France.

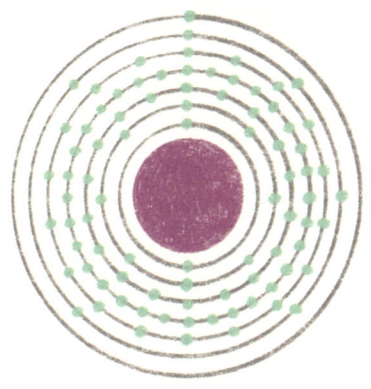

We now know that francium is the second-rarest element in nature. It is incredibly difficult to find, and only lasts a very short time before turning into other elements. Chemists had been searching for an element like francium for decades, to fill position 87 in the **periodic table**.

**87 Fr**

## A BRILLIANT CAREER

Marguerite was awarded money to finally go to university. She studied hard and became a very successful scientist. She continued studying francium, hoping it could help spot cancer early and save people's lives.

Sadly, unlike some other radioactive elements, francium isn't useful in this way. However, like other radioactive elements, it is very dangerous and can actually cause cancer – as it did with Marguerite herself.

Although francium may not save lives, it is still used in important scientific research today. Marguerite was nominated for the **Nobel Prize** in **Chemistry** five times, and was the first woman elected to the famous Paris Academy of Sciences.

*The periodic table organises the chemical elements. Turn to pages 60–61 to make your own table.*

*Scientists estimate there is less than 1 kilogram of francium in the Earth's crust at any one time. That's the weight of an average pineapple!*

29

# BIBHA CHOWDHURI
## (INDIA, 1913–1991)

Bibha was born at a time when most people didn't take women's education seriously. However, her parents believed it was very important for girls to get a good education.

When Bibha finished school, she studied **physics** at university. She then stayed on to complete a more advanced degree in physics, and was the only woman in her class.

After university, Bibha applied to be a research student at a special scientific centre. The male scientist who ran the research group didn't believe women should do this work. Bibha didn't give up and eventually convinced him to change his mind.

### SUBATOMIC PARTICLES

Everything in the universe (except energy) is made of tiny **atoms**. These atoms are made of even tinier subatomic particles, such as electrons, protons and neutrons.

### RESEARCHING RAYS

Bibha studied **subatomic particles** by examining **cosmic rays** from outer space in a special device called a cloud chamber. The particles leave trails as they pass through certain chemicals.

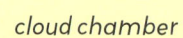

cloud chamber

*Bibha had to ride a mule high up into the Himalayan Mountains to set up cloud chamber experiments.*

### DISCOVERING THE PION

The particle trails disappeared very quickly but Bibha worked out how to capture their image on photographic plates. She discovered a subatomic particle that nobody had ever seen before. It was later named the pion, and it affects how the particles in an atom stay together.

### A MISSED OPPORTUNITY

During the Second World War, Bibha had to stop her research because she couldn't get the photographic plates she needed. A few years later, a scientist in England also found the pion, and won a **Nobel Prize**.

Bibha was still a successful physicist who spent her life doing important research. After she died, she was forgotten for a long time, but more people are now discovering her brilliant work. In 2019, a star was named after her!

# ROSALIND FRANKLIN
(ENGLAND, 1920–1958)

Rosalind was born in West London, into a close Jewish family. She always loved science, and went to university to study **chemistry**. During the Second World War, Rosalind volunteered as an air raid warden, helping to keep people safe during bombings.

She also took a job researching the chemistry of **coal**, to help the British government better understand how to use this fuel cheaply and safely. From this research, which Rosalind later used for her **PhD**, she learned about how the tiny holes in coal help it release energy and different substances when it is burned. She was fascinated!

After the war, Rosalind got a research job in Paris. She learned how to use **X-ray crystallography** to show how **molecules** are arranged in different materials.

## X-RAY CRYSTALLOGRAPHY

This technique involves shining an **X-ray** through a crystal to reveal its structure. Dots left on photographic paper show where the crystal's **atoms** are. Scientists then use maths to work out the substance's 3D structure.

*X-ray beam*

## PHOTOGRAPH 51

While working on a research project in England, Rosalind used X-ray crystallography to study the structure of **DNA** inside cells. One day, Rosalind took an image known as Photograph 51. It has been called 'the most important photograph in the world'.

## DNA

DNA is a molecule found inside the cells of living things. It contains information to make each living thing look and function in certain ways.

Photograph 51 showed that DNA has an unusual shape. It is made of two strands of genetic information, curled around each other in a spiral. This curled ladder shape is called a 'double helix', and it splits apart to make copies of itself.

Rosalind's work was central to the model of DNA's structure made by scientists James Watson and Francis Crick. Very sadly, Rosalind died young and so could not share their **Nobel Prize** for this discovery. However, her work is still helping to save lives and push science forwards today.

*photographic paper*

*crystal*

*Rosalind also did important work on the structure of viruses. Many people believe she would have won a Nobel Prize for this, if she had lived longer.*

33

# STEPHANIE KWOLEK
(USA, 1923–2014)

When Stephanie was little, her father taught her about nature and science. After he died, her mother worked making clothes, and inspired Stephanie's love of fabrics and design.

Stephanie went to university to study **chemistry**. She then took a job as a **researcher** for a **company** that made synthetic fibres – materials made from chemicals, rather than natural materials such as cotton.

## MAKING MATERIALS

One day, while Stephanie was trying to find a lightweight material to strengthen car tyres, she noticed something odd. One of the **polymer** fibre mixtures she'd made was cloudy, instead of clear as she'd expected.

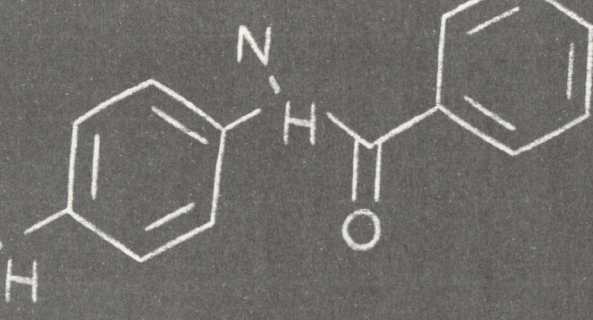

Instead of seeing this as a mistake, Stephanie tested the mixture to see what kind of material it would make. She passed it through tiny holes in a device called a spinneret. This turned the mixture into a fibre, and its **molecules** now lined up in a neatly ordered structure.

### POLYMER

All materials are made of tiny particles called **atoms**. These atoms usually join up together in groups called molecules. A polymer is a large molecule made up of monomers – small, repeating blocks of atoms.

## A FASCINATING FABRIC

Stephanie had created a light, flexible, **fire-resistant** fibre that was five times stronger than steel! It was woven into a fabric called Kevlar®, which can be used as protective, bullet-proof **armour**. Kevlar® is also used to make planes, boats, cars, skis, tyres, shoes and much more.

Stephanie never made money from her brilliant, life-saving invention. She signed the **patent** over to the company where she worked, which has made millions of dollars from Kevlar®. However, Stephanie won many awards, successfully applied for 17 other patents, and worked hard encouraging women to become scientists.

Stephanie invented the Nylon Rope Trick, a way to make a synthetic fibre called nylon. It is now a popular classroom experiment. Ask an adult to find you a video of it online!

# VERA RUBIN
(USA, 1928–2016)

From a very young age, Vera was fascinated by the night sky. She joined her local **astronomy** club, and her parents helped her build a telescope from a **lens** and a cardboard tube.

Vera studied astronomy and **physics** at university, and later taught students herself. She then worked as a **researcher**, using telescopes to study **galaxies**.

*We can use telescopes to see far-off stars.*

### GALAXIES

A galaxy is a system of millions or billions of stars, and their solar systems, plus huge amounts of gas and dust. A solar system includes objects that move around a star, such as planets and moons.

## STRUGGLES AND STARS

In Vera's time, some men didn't think women should study astronomy. They refused to let women join certain university programs or to use special equipment at **observatories**. Vera made fun of these silly, angry men, but also had to unfairly struggle to do the work she loved.

One day, Vera noticed something strange in her **observations**. The stars at the outer edges of one type of galaxy were moving just as fast as the stars close to its centre. To keep hold of such speedy, far-off stars, the galaxy must have a much stronger force of **gravity** than it should for its size. Vera realised that there is something in space that has gravity but is not visible.

# TU YOUYOU
## (CHINA, BORN 1930)

When Youyou was a teenager, a disease called **tuberculosis** made her so ill she missed two years of school. This difficult time inspired her to study medicine, so she could find cures to help other people.

At university, Youyou studied pharmacology – the science of how medicines work. She then got a research job at the Academy of Chinese Traditional Medicine.

A few years later, the **Chinese Cultural Revolution** began. Many scientists were sent away from their homes and not allowed to work. Youyou's husband was sent away but she was allowed to keep working.

### SEARCHING FOR A CURE

Youyou was put in charge of Mission 523, a secret plan to find a cure for **malaria**, a very serious illness. She was sent to a Chinese island to study how malaria affected people's health. She couldn't see her two young daughters for around two years.

*Malaria is typically spread through bites from female Anopheles mosquitos.*

38

By the time Youyou started searching for a malaria cure, scientists around the world had tested over 240,000 substances – with no success. She and her team researched hundreds of plants used in Chinese medicines. They even looked through ancient Chinese books for clues to possible cures.

In one book, Youyou discovered instructions from around 1,500 years ago for using the sweet wormwood plant to treat malaria. She **extracted** a drug called artemisinin from this plant, and tested it on herself to show it was safe.

## A LIFE-SAVING RESULT

Artemisinin was a very successful malaria treatment, but it took years more study and testing before it was officially approved. In the 1980s, the importance of artemisinin was finally shared with the world.

Youyou's discovery has saved millions of people's lives. She has won many awards for her amazing work, and is the only woman from China who has been awarded a **Nobel Prize**.

*Youyou's name was inspired by a poem that describes the sound deer make when they eat the sweet wormwood plant!*

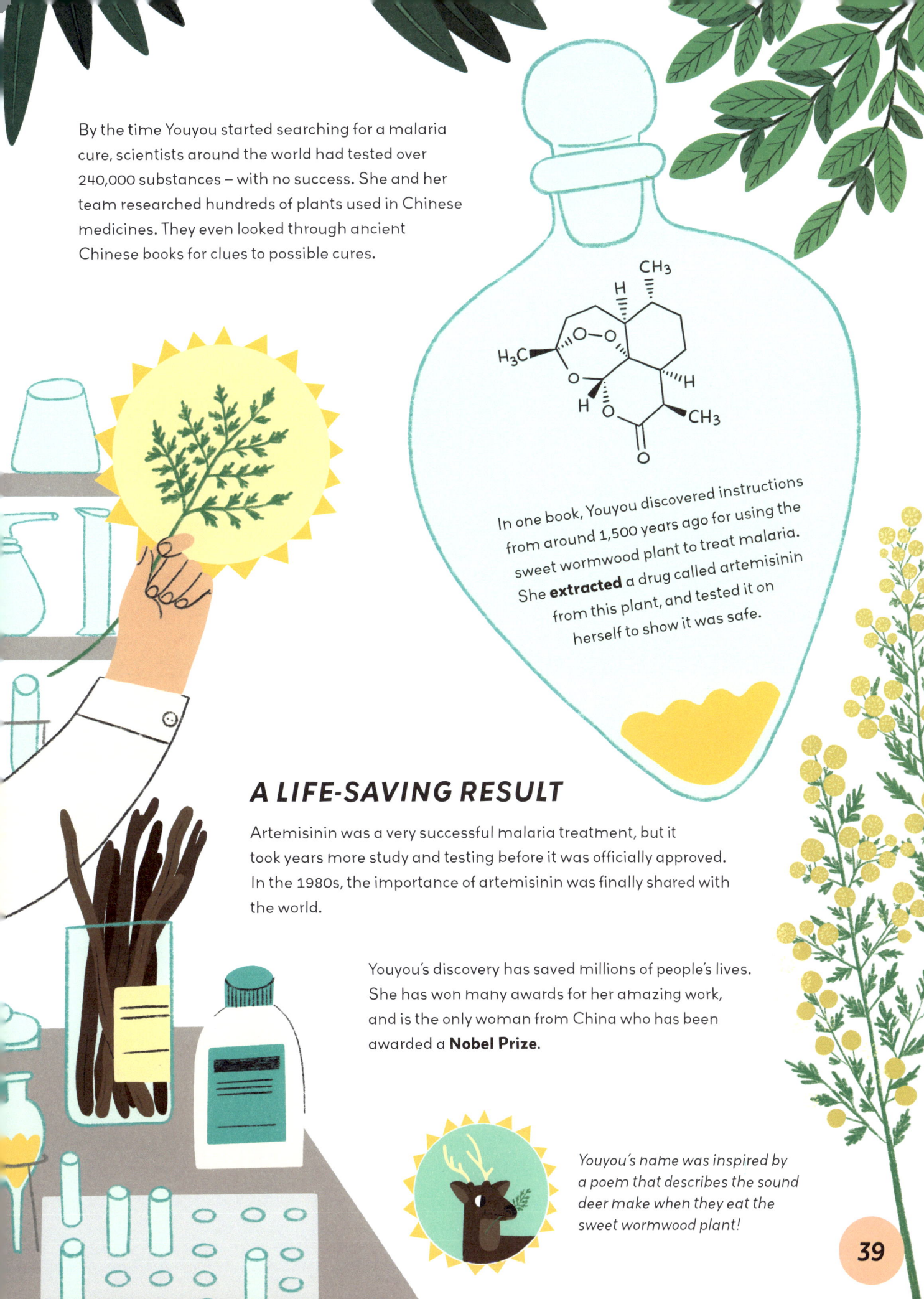

# ROSELI OCAMPO-FRIEDMANN
## (PHILIPPINES, 1937–2005)

Roseli was always fascinated by plants. She studied botany – the science of plants – at university in her hometown of Manila, in the Philippines. She continued her studies in Israel and the USA, researching different types of **microorganisms**.

Roseli discovered her great talent for helping microorganisms survive and grow in the **laboratory**. She and her husband travelled around the world, searching for new microorganisms in places where people thought it was impossible for anything to live.

*bacteria*

### MICROORGANISMS

Microorganisms are very tiny living creatures. They are so small that you cannot see them without a **microscope**. They include bacteria, viruses, algae and fungi.

*viruses*

*fungi*

*algae*

### EXTREME LIVING

Roseli found many microorganisms that can survive in very difficult conditions, such as extreme heat or cold. They are called extremophiles. On a trip to Antarctica, she discovered new microorganisms living inside frozen rocks!

These creatures were named cryptoendoliths, which means 'hidden inside rocks'. Roseli helped them stay alive and **reproduce** in her laboratory, and studied them closely.

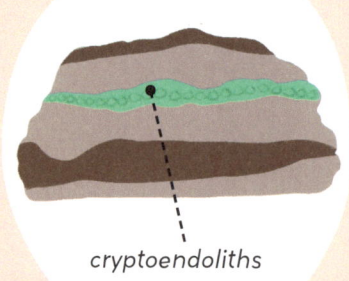

*cryptoendoliths*

Roseli eventually became a **professor** at Florida State University, and received awards for her research work. She continued her research trips, including studying bacteria living in the permafrost (permanently frozen ground).

## LIFE ON MARS?

Roseli is thought to have found more than 1,000 different extremophiles all over Earth, but her work is also important for scientists looking for life on other planets! The conditions on Mars are similar to those where Roseli found microorganisms in Antarctica, so her research has offered scientists some clues.

NASA has also used Roseli's work to help them explore whether humans could ever live and grow food on Mars.

*Extremophiles have been found living inside clouds, active volcanoes and toxic acid lakes.*

*A mountain in Antarctica is named Friedmann Peak in Roseli's honour.*

# OMOWUNMI SADIK
## (NIGERIA, BORN 1964)

Omowunmi grew up in a family that loved to learn. Her father taught her the basics of science, and she studied with her brothers and sisters at home. She loved to ask questions, especially about science.

Omowunmi studied **chemistry** at university, and after getting her **PhD** she eventually became a **professor** in the USA. She was fascinated by **biosensors** and learned so much about them that she ended up inventing her own new ones! Over the years, Omowunmi has invented biosensors to help with all sorts of problems.

### BIOSENSOR

'Biosensor' is short for 'biological sensor'. It is a small device that uses a substance taken from a living thing to sense a certain chemical. It then sends its findings to an electronic device.

### BRILLIANT BIOSENSORS

One of Omowunmi's biosensors checks whether fresh food contains *E. coli* bacteria, which can make people very ill. Another can find tiny amounts of drugs and explosives on the surface of objects.

Omowunmi has invented a biosensor that can detect a virus called HIV in minutes, rather than the three to four days needed for standard tests. She's also developed similar devices to help doctors test for Covid-19 in places far from testing **laboratories**.

Omowunmi has used biosensors to help small farmers find **microorganisms** that stop crops growing properly. All the farmer has to do is put some soil on a special paper strip, and they can see the results on their mobile phone.

*information*

*soil*

*biosensor*

*paper*

## SUSTAINABLE SCIENCE

Omowunmi has developed ways to remove toxic pollution from the environment, and to **recycle** it into useful, safer products. She has also studied how the teeny particles used in **nanotechnology** might affect the long-term health of humans and the environment.

Companies now use nanoparticles for everything from making tennis balls bouncier to stopping socks getting smelly. Omowunmi leads a global organisation focused on using nanotechnology safely and sustainably.

*Some of Omowunmi's biosensors are known as 'robotic noses'. Turn to page 58 to play your own 'smell test' game!*

As one of the world's most respected scientists, Omowunmi has won many awards and honours. She continues to research, invent and teach, helping and inspiring people all over the world.

43

# NZAMBI MATEE
(KENYA, BORN 1992)

Nzambi studied **physics** at university, and learned how different materials can be used. She also taught herself **engineering**, which involves using science to solve practical problems. When she saw the mess caused by plastic pollution near her home, she wanted to help clean it up.

Nzambi quit her job and started to experiment with plastic waste in her mother's backyard. Even though her neighbours complained about the noise, she worked very hard and started mixing plastic with sand to see if she could create something useful.

## BUILDING BLOCKS

Nzambi discovered that certain plastics mix together very well, and some of these mixtures can be used to make paving blocks. She won a **scholarship** to go to the USA and use special **laboratories** to work on her invention.

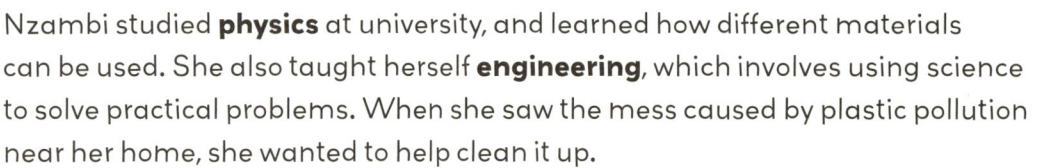

The blocks Nzambi invented are light, affordable, and up to seven times stronger than **concrete**. They are perfect for building hard-wearing paths that don't get muddy when it rains. Nzambi also designed and built machines to create these blocks.

## SAVING THE PLANET

Nzambi set up a **company** to make and sell the paving blocks. She buys plastic from **recycling** centres or collects it from companies that usually pay to have it taken away. The waste material is mixed with sand, then heated and pressed into brick shapes. Her company makes more than 1,000 paving blocks each day.

People want to buy more paving blocks than her **factory** can produce, so she is growing her business to meet this huge demand. Nzambi has won many prizes for her work, and in 2020 the **United Nations** Environment Programme named her a Young Champion of the Earth. She hopes that in the future she can create recycled plastic bricks for building houses.

*It takes 700 years for plastics in landfills to start breaking down.*

*Less than 10% of the world's plastic is currently recycled.*

*Can you recycle rubbish into something useful? Turn to page 59 to make your own invention inspired by Nzambi.*

45

# DRAW A REALISTIC FLOWER

Try creating a scientific drawing, like Maria Sybilla Merian! She made discoveries about the natural world by looking carefully at her specimens and drawing exactly what she saw. What might you discover?

### YOU WILL NEED

two pieces of paper (or one sheet divided into two)

pencil

colouring pencils

a flower – ask an adult's permission before you pick any flowers, and try to find one that has already fallen on the ground; always wash your hands after touching it

flat-tip tweezers

**1.**

First, choose a nice view of your flower. Set it up in front of you in that position.

**2.**

Look at the flower closely, and pick out the shapes that make it up. Draw those shapes to create an outline of the flower.

**3.**

Next, look at the textures and patterns on the flower. Add those to your drawing, either with a pencil or in colour. Try mixing colours, adding one shade on top of another, to get the colours you see on the flower.

**4.**

When you are happy with your drawing, try dissecting (taking apart) the flower to reveal its insides. Use the tweezers and your fingers to carefully tease apart the petals – or take some off, if necessary, to see inside.

**5.**

*stigma*

*anther*

*filament*

*style*

*petal*

*ovary*

*sepal*

*ovule*

*stalk/stem*

Look carefully at the inside of the flower. Try to spot parts such as the stigma, style, ovary, anther, filament and stem. Remember that these can look different in different flowers.

**6.**

Repeat steps 1 to 3, this time drawing the inside of the flower. Add labels to your drawing for the parts (stigma, style and so on) that you've identified.

### WHAT'S THE SCIENCE?

You might not think drawing is a science but detailed, accurate pictures are very important for helping scientists understand the natural world. By capturing her discoveries in realistic drawings, Maria proved how natural processes, such as insect **life cycles**, really worked.

### TRY IT AGAIN – DIFFERENTLY!

What else from nature could you draw an inside-and-outside picture of? Perhaps a seed pod, a piece of fruit or a vegetable? Ask an adult for permission before you take or open up anything!

# MAKE YOUR OWN FOSSIL

**Fossils**, like the ones that Mary Anning discovered, take thousands of years to form. But you can try creating a quicker version in just a few days! Instead of waiting for mud to turn into rock, you can use salt dough to make solid casts.

### YOU WILL NEED

250g plain flour (plus extra for kneading)

250g table salt

125ml warm water

bowl for mixing

paper cup or small card box

plasticine or sticky tack

objects with interesting shapes to 'fossilise', such as seashells or plastic toys

an adult to supervise

**1.** First, make the salt dough. Mix the flour and salt together. Then slowly add the warm water, stirring until it becomes a soft dough.

**2.** Sprinkle a bit of flour on a flat surface. Knead the dough on this floured surface for several minutes, until it becomes springy and elastic.

**5.** Leave the paper cup or card box somewhere warm for a few days. Ask an adult to help you find a safe place for it that small children and pets can't reach.

**6.** When the salt dough feels hard, cut open the cup or box and carefully pull apart the plasticine and dough. This should reveal salt dough 'fossils' of your object!

48

**3.** Press the plasticine or sticky tack into the paper cup or card box. Then carefully push the object into the plasticine, before taking it out again. Set it aside.

**4.** The impression (shape) of the object should now be marked into the plasticine. Add a layer of salt dough over the top of the plasticine, carefully pressing it into the impression left in the plasticine.

### TRY IT AGAIN – DIFFERENTLY!

Can you use the plasticine and salt dough to capture a 'trace', such as a footprint or handprint? Trace fossils are another important type of fossil, which also help us understand more about dinosaurs and other living things from the past.

### WHAT'S THE SCIENCE?

Like a buried dinosaur, the object made a printed shape in the 'mud' (plasticine). This shape is called a cast. You filled the cast with salt dough, which hardened as it dried out. This left a 3D fossil of the object in the 'rock' (salt dough).

49

# DESIGN A BRIDGE

Have a go at designing a bridge, like Sarah Guppy did.
See how strong a bridge you can build, using just spaghetti and marshmallows!

## YOU WILL NEED

spaghetti (uncooked!)

packet of mini marshmallows

two books, or other solid objects, to build a bridge between

small orange or other similar-sized piece of fruit

ask an adult before you begin

Set up both sides of your bridge, with a gap in between. You could use two books, chairs or other solid objects of roughly the same height.

Use the mini marshmallows to connect the spaghetti sticks in any way you can think of. You can snap the spaghetti sticks into the lengths you need.

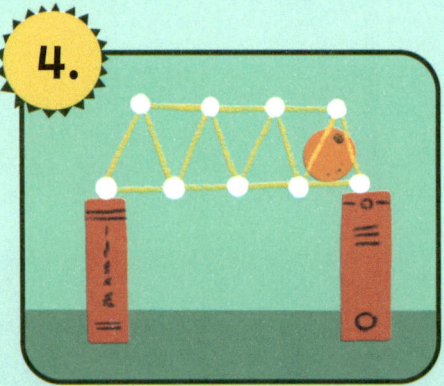

When your bridge is complete, place it over the gap and test out its strength! Can it hold the orange's weight? Start by placing the fruit close to one edge.

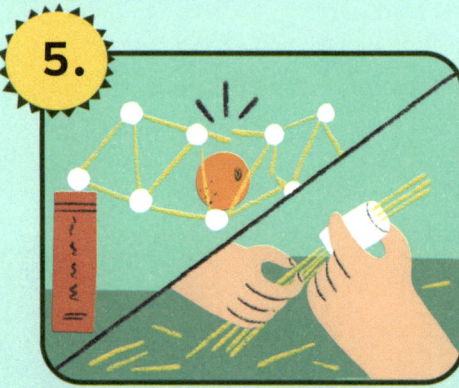

If your bridge holds, move the orange towards its middle. If it moves around or breaks under the weight of the orange, try using more strands of spaghetti to make it stronger.

### 3.

Use your imagination and creativity to plan a bridge design. Make sure it's long enough to reach all the way across the gap.

### 6.

When you've finished, you could avoid food waste by eating the spaghetti and mini marshmallows. Ask an adult to cook the spaghetti first!

## WHAT'S THE SCIENCE?

You should find that using lots of spaghetti strands bunched together makes your bridge stronger. This is similar to how **engineers** create strong cables that help hold up heavy bridges. Each cable contains lots of steel wire strands inside a plastic coating.

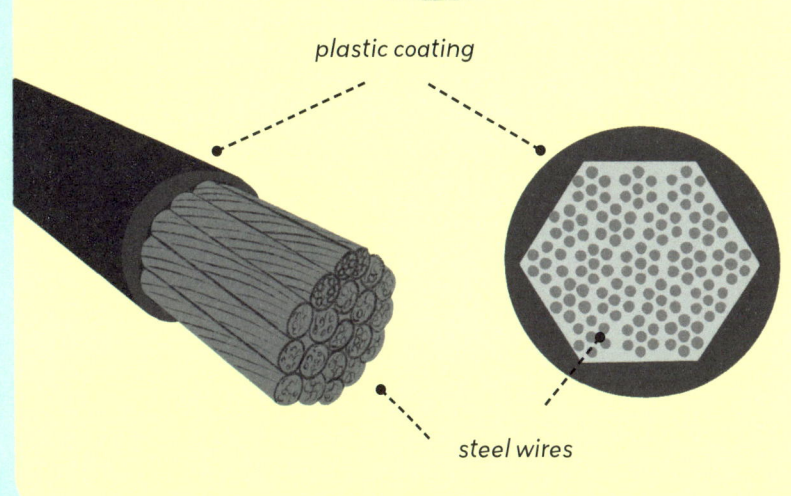

*plastic coating*

*steel wires*

## DO IT AGAIN – DIFFERENTLY!

What changes could you try out to make a stronger bridge? Think about different materials you could use, and different ways of building it.

## FUN WITH FRIENDS

Challenge a friend to see who can create the strongest bridge! Try putting heavier and heavier pieces of fruit on each of your bridges. Whose bridge lasts the longest before it breaks?!

# MAKE PERFUME – FOR YOUR HOME!

You can make **perfume** just like Tapputi, using simple kitchen equipment and an adult to help you. Use it to make your home smell amazing!

## YOU WILL NEED

scented, non-toxic flowers and herbs such as lavender, rose, mint, rosemary, sage – *be mindful of allergies or skin reactions*

water

empty jam jar

pestle and mortar or rolling pin (optional)

filter paper – try coffee filters

funnel

empty spray bottle

an adult to supervise

### 1.

Ask an adult to help you pick a few scented petals or leaves from a plant they know is safe for you to touch. Rub them gently, then smell your fingertips to see which scents you like best.

### 2.

Choose one flower/herb, or a combination of two or more. Gently rub your pestle or rolling pin over the petals/leaves to squash them a bit. If you don't have either, squeeze and tear them.

### 3.

Put the bashed leaves and petals into the jam jar and top up with water. Leave the jar in a warm place overnight.

### 4.

The next day, put some filter paper into the funnel (or use a tea strainer). Pour your perfume liquid through it into the spray bottle. Spray around your home and enjoy!

## TRY IT AGAIN – DIFFERENTLY!

Try different combinations of herbs, petals and fruit peel. Which work best? Notice how long the perfume lasts before it starts smelling 'off'!

## WHAT'S THE SCIENCE?

When you bashed the leaves, the cells broke open, releasing the scent inside the leaves. While the leaves sat in the water, the scent dissolved into the water.

## WARNING!

Never eat or drink the plant extracts as they could make you very unwell. Some common plants and trees are poisonous if you eat parts of them.

# SET OFF A CHAIN REACTION

Lise Meitner discovered that when you split a **uranium atom** apart, the neutrons (tiny particles) that explode out of the atom go on to split apart other nearby atoms. Have a go at setting off a chain reaction – with dominoes or blocks rather than uranium atoms!

### 1.

Stand one block on its short end. Stand two more blocks the same way up, quite close to the first block, to make a triangle shape. The idea is that when the first block falls, it will knock over the two blocks in front of it.

### 2.

Stand four more blocks on end, two each in front of the second and third blocks. You'll notice that you need to fan the block outwards as you go, to have enough space.

### 3.

Keep going until you run out of space, or run out of blocks.

### 4.

Lightly tap the first block and watch what happens! If the blocks don't all fall down, it's probably because some of them were too far apart. Move them closer together and try again.

### YOU WILL NEED

box of dominoes or blocks – rectangular ones work best

a hard surface, such as a floor or table

### WHAT'S THE SCIENCE?

When one action affects more than one object, and that same pattern continues at each step, that one small action ends up having a huge effect. This is what happens with energy release in **nuclear fission**.

### TRY IT AGAIN – DIFFERENTLY

Instead of knocking over lots of blocks of the same size, try knocking over a chain of objects that get bigger at each step. What is the biggest object that you can knock over at the very end of the chain?

# MAKE A PAPER BAG

Margaret Knight designed a machine to fold paper into a bag with a flat bottom. Try making this paper bag by hand!

### YOU WILL NEED

sheet of A4 paper – thick, coloured paper works well

glue stick or sticky tape

two pieces of ribbon or string (optional)

ask an adult before you begin

**1.** Place your sheet of paper on a flat surface. Fold the two short edges of the paper into (roughly) the middle, until they overlap. Tape or glue them together.

**2.** Fold up the open end closest to you, to about a third of the way up the long sides. Press it down along the crease, then unfold it again.

**5.** Gently pop out the bottom of your bag, and carefully squeeze the long sides to open up the inside.

**6.** If you like, add handles to make the bag easier to carry. Try taping a looped piece of ribbon or string on each side inside the bag's open top.

**3.** Push your fingers into the sides of the open end closest to you. Open this fold, flattening the corners to make triangular shapes. Press down along the creases.

**4.** Let's finish making the bottom of your bag! Fold each side of the bottom towards the middle, until they overlap. Tape or glue them together.

### TRY IT AGAIN – DIFFERENTLY!

Experiment with different sizes of folds for the bottom of the bag. Try making the bottom fold a quarter or a half of the bag's length. Which feels sturdiest? You could try using different sizes of paper sheets too.

How about re-using wrapping paper or perhaps even making your own design by painting on the paper first?

### WHAT'S THE SCIENCE?

You have turned a sheet of paper into a paper bag just by folding and sticking it together in a certain way. Margaret designed a machine to copy this process by first working out which actions it would need to do, in which order. She then built moving parts to carry out each of these actions in turn.

# MAKE A POND VIEWER

Want to get a closer look at the creatures living in your local pond or rockpool? Get inspired by Jeanne Villepreux-Power and make your own pond viewer!

## YOU WILL NEED

empty plastic milk bottle or fizzy drink bottle – the bigger the better

round-ended scissors

duct tape

rubber bands

cling film

an adult to supervise

**1.** Ask an adult to help cut off the bottom of the bottle.

**2.** Cover the open bottom of the bottle with a large piece of cling film. Leave plenty of spare cling film around the sides.

**5.** Find a place where you can safely put the pond viewer into the water, without risk of falling in yourself. Lower it in gently, keeping the side with the cling film at the top.

**6.** Wait for your pond viewer to fill with water – and, hopefully, wild creatures! When you've finished looking at them, gently lift the pond viewer out of the water and make sure no creatures are left stuck inside.

**WARNING!**
Make sure an adult is with you when you are near water! Wash your hands after using your pond viewer.

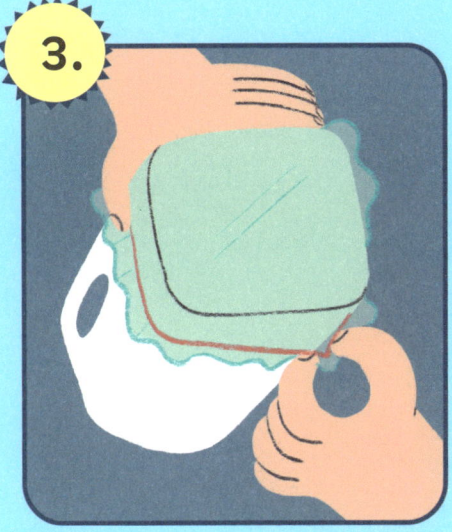

Use a rubber band or two to secure the cling film in place. Gently pull the cling film tight so that it lies straight and smooth over the opening.

Use the duct tape to seal the cling film around the sides of the bottle, so that no water can get in or out.

### WHAT'S THE SCIENCE?

Just like Jeanne's aquarium, the pond viewer is a window into a watery world we don't normally get to see. You can see wild creatures up close and watch how they behave in their natural habitat.

### TRY IT AGAIN – DIFFERENTLY!

Try out the same method with different materials. Which bottles give you the best view? If you stretch the cling film tighter, is it easier to see through?

57

# PLAY THE 'SMELL TEST' GAME

Omowunmi Sadik has invented biosensors that are sometimes known as 'robotic noses'. But how good a detective is your nose? Play this smelling game to find out!

## YOU WILL NEED

five small bowls (or plates)

five strong-smelling items from the kitchen or bathroom – for example, herbs, spices, soaps and shampoo – that are safe for all players to handle; an adult will need to choose these items, being aware of players' allergies and so on

one or more friends or family members to play with, including an adult to be the tester

blindfold for each person

### 1.

Ask the adult tester to put a little of each chosen item into its own bowl. They need to do this without any of the players (including you!) seeing the items.

### 2.

Each player puts on their blindfold. The tester then carefully brings the 'smell test' bowls into the room.

### 3.

The tester passes around one bowl at a time. They ask each player to smell it and guess what the item is. Players can whisper their answers to the tester, to avoid influencing other players' guesses.

### 4.

How many smells can each player guess correctly? The player with the most correct answers is the winner!

## WHAT'S THE SCIENCE?

When you smell something, you're sensing tiny particles – called odorants – that you've breathed in. When odorants reach the back of your nose, special cells detect them and send messages to your brain. It then works out what you're smelling!

## TRY IT AGAIN – DIFFERENTLY!

If you play a second round, try to include a smell that at least one person didn't guess correctly the first time. Will they remember it and get it right this time?

## WARNING!

Make sure that you only sniff the items and don't taste any of them! Some common items could make you very unwell.

58

# BUILD A RECYCLED TOY HOUSE

Nzambi Matee worked out how to turn plastic rubbish into useful paving blocks. Can you make a house for a toy, using the contents of your **recycling** bin?

## YOU WILL NEED

- clean, safe materials from your recycling bin – paper, cardboard, empty plastic bottles, boxes, cartons and so on; ask an adult to help you gather these
- sticky tape
- round-ended scissors
- a toy – maybe an old favourite teddy or action figure
- crayons or felt-tip pens

### 1.

Write a list of the different parts that can make up a house. Try looking at pictures of different houses from around the world.

### 2.

Look through the materials you've collected. Which material might work best for each part of your house? For example, if you want to make a window, you could use a piece of clear plastic.

### 3.

Draw a rough design for your toy house. Don't worry about it looking perfect – it's just a plan to help you build it. Label each part of the house with the material you're using to make it.

### 4.

Cut and stick together the materials to build your toy house. Make sure it's big enough for your toy to fit in comfortably!

## WHAT'S THE SCIENCE?

When you chose which materials to use for your house, you thought about their properties – that is, how strong, lightweight or see-through they are. Nzambi similarly chose materials with suitable properties to create tough, lightweight blocks.

### TRY IT AGAIN – DIFFERENTLY

Build something different for a toy – maybe a car, a boat or a bed. Draw a design and choose which clean, safe materials from the recycling bin have the most suitable properties for each part.

# DESIGN YOUR OWN PERIODIC TABLE

The periodic table organises all known **elements** according to their size and how they look and behave. When Marguerite Perey discovered francium, it filled an empty place in the table (number 87). Can you create a periodic table for your favourite belongings?

### YOU WILL NEED

a selection of your favourite belongings – around 20–30 toys, books, art supplies and any other special things that are safe to handle

a clear, clean space on the floor or table

pencil and paper

colouring pencils or felt-tip pens

ruler (optional)

ask an adult before you begin

**1.** First, group your belongings by colour. Put all the red objects into one pile, all the blue objects into another pile, and so on. Arrange all the piles in a horizontal line, in any order you like.

**2.** Now sort each colour group by size. Arrange the objects in each group into a vertical line, with the smallest object at the top and the biggest at the bottom. You should now have a 'grid' of objects arranged by colour and size.

**3.** When you are happy with your grid, you can draw it on your paper. You can use a ruler to draw in the horizontal and vertical rows, creating a box for each object. In each box, draw or write the name of the object. Colour in and decorate your periodic table – and hang it on the wall!

### WHAT'S THE SCIENCE?

Organising your favourite belongings by colour and size encourages you to pay closer attention to them. You might notice, for instance, that your biggest green object is smaller than your smallest blue object! Lots of scientists try to organise the different things that they study, to help them notice patterns, gaps or anything else interesting.

### TRY IT AGAIN – DIFFERENTLY!

Create a new periodic table by reorganising your favourite belongings in a different way. Maybe you could group them according to the type of object (toy, book and so on), and put them in order from lightest to heaviest?

# THE PERIODIC TABLE

*Turn the book on its side to read the elements of the periodic table!*

| Period \ Group | 1 | 2 | 3 | 4 | 5 | 6 | 7 | 8 | 9 | 10 | 11 | 12 | 13 | 14 | 15 | 16 | 17 | 18 |
|---|---|---|---|---|---|---|---|---|---|---|---|---|---|---|---|---|---|---|
| 1 | 1 H | | | | | | | | | | | | | | | | | 2 He |
| 2 | 3 Li | 4 Be | | | | | | | | | | | 5 B | 6 C | 7 N | 8 O | 9 F | 10 Ne |
| 3 | 11 Na | 12 Mg | | | | | | | | | | | 13 Al | 14 Si | 15 P | 16 S | 17 Cl | 18 Ar |
| 4 | 19 K | 20 Ca | 21 Sc | 22 Ti | 23 V | 24 Cr | 25 Mn | 26 Fe | 27 Co | 28 Ni | 29 Cu | 30 Zn | 31 Ga | 32 Ge | 33 As | 34 Se | 35 Br | 36 Kr |
| 5 | 37 Rb | 38 Sr | 39 Y | 40 Zr | 41 Nb | 42 Mo | 43 Tc | 44 Ru | 45 Rh | 46 Pd | 47 Ag | 48 Cd | 49 In | 50 Sn | 51 Sb | 52 Te | 53 I | 54 Xe |
| 6 | 55 Cs | 56 Ba | 71 Lu | 72 Hf | 73 Ta | 74 W | 75 Re | 76 Os | 77 Ir | 78 Pt | 79 Au | 80 Hg | 81 Tl | 82 Pb | 83 Bi | 84 Po | 85 At | 86 Rn |
| 7 | 87 Fr | 88 Ra | 103 Lr | 104 Rf | 105 Db | 106 Sg | 107 Bh | 108 Hs | 109 Mt | 110 Ds | 111 Rg | 112 Cn | 113 Nh | 114 Fl | 115 Mc | 116 Lv | 117 Ts | 118 Og |
| 6 | | | 57 La | 58 Ce | 59 Pr | 60 Nd | 61 Pm | 62 Sm | 63 Eu | 64 Gd | 65 Tb | 66 Dy | 67 Ho | 68 Er | 69 Tm | 70 Yb | | |
| 7 | | | 89 Ac | 90 Th | 91 Pa | 92 U | 93 Np | 94 Pu | 95 Am | 96 Cm | 97 Bk | 98 Cf | 99 Es | 100 Fm | 101 Md | 102 No | | |

*Turn to page 72 for a list of the 118 elements in the periodic table, arranged by atomic number!*

# MEASURE SOMETHING YOU CAN'T SEE

Vera Rubin discovered **dark matter**, using maths to prove that it must exist even though we can't see it. Measuring something you can't see sounds difficult, but it's not impossible. Wind is something we can't see – so how can we measure it?

### YOU WILL NEED

an adult to supervise

the Beaufort scale (opposite page)

a safe place to walk outside, or an open window to look out of

**1.** Take this book with you out on a walk, or over to an open window. Make sure an adult is with you either way.

**2.** Look around for any trees, water or smoke, and study how the wind is moving (or not moving) them. Notice how the wind feels on your skin, if you can feel it at all.

**3.** Read through the Beaufort scale on the opposite page. Decide which grade best describes the wind right now. Ask the person you're with which grade they have chosen.

Try repeating the steps above once a week for a few weeks. Write down the grade for each day in a notebook, to look back on and compare.

### WHAT'S THE SCIENCE?

Scientists often study things that can't be seen and measured in a traditional way. They have to think more creatively about how to do this, such as looking at the effect that something has on its surroundings. This is what the Beaufort scale does for the wind.

### TRY IT AGAIN – DIFFERENTLY

Create your own fun scale to measure something that you can't see! Our world is full of things that we hear, feel, smell and taste. Think about what you want to test, and how you could measure it. Ask an adult to help you do this safely, and make a scale so others can have a go too!

| Beaufort grade | Observation | Wind speed (km per hour) |
|---|---|---|
| **0** Calm and still | The water surface is still and smooth. Smoke rises straight up. | 0 – 1 |
| **1** Light air | Small ripples on the water. Smoke drifts in the direction of the wind. | 1 – 5 |
| **2** Light breeze | Can feel the wind on your skin. Leaves rustle in trees. | 6 – 11 |
| **3** Gentle breeze | Leaves and small twigs constantly moving in trees and bushes. | 12 – 19 |
| **4** Moderate breeze | Small branches blow in the wind. Dust and light litter blow about. | 20 – 29 |
| **5** Fresh breeze | Small trees sway in the wind, and medium-sized branches move about. | 30 – 39 |
| **6** Strong breeze | Large trees move in the wind, and it's hard to hold up an umbrella. | 40 – 50 |
| **7** Near gale | Whole trees are moving. It's hard to walk against the wind. | 51 – 61 |
| **8** Gale | Twigs break off from the tree. Cars struggle to drive straight. | 62 – 74 |
| **9** Severe gale | Small trees blow over, and branches can fall off trees. | 75 – 87 |
| **10** Storm | Roof tiles fall off. Trees uprooted. | 88 – 101 |
| **11** Violent storm | Lots of damage to plants. Some damage to buildings. | 102 – 117 |
| **12** Hurricane | Windows broken, sheds and barns damaged. Objects flying in the wind. | 118 or more |

# MAKE A LAVA LAMP

Inge Lehmann discovered that Earth has a solid inner core, surrounded by a fluid outer core. Inside the fluid core, the liquid hot rock slowly moves up and down. Can you try to make the oil in this lava lamp do the same?

## YOU WILL NEED

empty, clear plastic bottle

water

cooking oil

food colouring

salt

torch (optional)

an adult to supervise

**1.** Fill the bottle one-third full with water.

**2.** Pour cooking oil into the bottle, leaving at least 3 cm space at the top. If this is too much oil for you to use, just make sure there's roughly three times as much water as oil.

**5.** If you have a torch, try shining it up through the bottom of the bottle to create a glowing lamp effect.

**6.** When the bubbles run out, you can start them off again by adding some more salt.

*The lava lamp was invented in 1963 by Edward Craven Walker. People loved watching its blobs of coloured wax rise and fall, and it was popular in homes through the 1960s and 1970s.*

**3.** Add some food colouring – just a few drops, so the water doesn't go too dark. Look closely at where the food colouring goes. Does it colour everything?

**4.** Next, add salt until you see something start to happen... What do you see?

### WHAT'S THE SCIENCE?

The oil floats on top of the water. The food colouring only mixes with the water. The 'lava lamp' bubbles are formed when the salt mixes with water, and the reaction gives off gas. The gas travels up through the oil, taking the coloured water with it. When the gas escapes, the coloured water part falls back down to the bottom.

### TRY IT AGAIN – DIFFERENTLY!

Try using warm water. Does the water temperature make a difference to how many bubbles there are, and/or how quickly they move?

# GLOSSARY

**Anatomy**
The science of how the different parts and systems of living things fit together and work.

**Armour**
Clothing worn to protect the body from getting hurt by weapons and fighting.

**Astronomy**
The study of everything in the universe outside the Earth's atmosphere. This includes stars and planets. A person who studies astronomy is called an astronomer.

**Atom**
The smallest whole particle of a chemical element. It can be broken open to release the even smaller particles inside.

**Chemistry**
The study of what things are made of and how those things work together. A person who studies chemistry is called a chemist.

**Chinese Cultural Revolution**
A time of huge political and social change in China, from 1966 to 1976.

**Company**
A business organisation.

**Compound**
A substance made up of two or more different elements.

**Concrete**
A strong, hard material made from rock, sand and water. It is often used for building.

**Cosmic rays**
High-energy subatomic particles that produce secondary particles, such as the pion, when they come to Earth from outer space.

**Cotton mill**
A building with machines to turn cotton (a soft, white plant fibre) into thread or cloth.

**Dark matter**
Anything in the universe that has mass but does not give off or reflect light. This means that we cannot see it.

**Distillation**
A process that turns liquid into a vapour, and then back to a liquid.

### Earthquakes
Movements of the ground caused by shifts in Earth's crust.

### Element
A basic substance that cannot be broken into any different parts.

### Engineer
A person who uses their knowledge of science to design and create machines and buildings.

### Enslaved
Describes a person who is a victim of slavery. They are considered to be the property of another person, and are forced to work for that person.

### Excavate
To uncover something, such as a fossil, by digging away the earth and other material that covers it.

### Expedition
A journey with a particular purpose, such as scientific research, typically taken by a group of people.

### Extract
To separate out and remove a certain substance from a material or mixture.

### Factory
A building where people and machines make things.

### Felt
A cloth made by pressing or rolling together a soft material such as wool.

### Filtered
When a liquid or gas is passed through a filter to remove unwanted elements.

### Fire-resistant
Something that is difficult to burn or set fire to.

### Fossil
The remains of a plant or animal that has been preserved in sand or mud for thousands, or even millions, of years and eventually turned to stone.

### Geology
The study of Earth's physical structure, and how this has changed over time. Geologists typically focus on studying rocks.

### Gravity
A pulling force where a large object pulls smaller things towards it.

**Industrial Revolution**
A period when technology and machines became an important part of the way people lived and worked. It began in Europe at the start of the 18th century.

**Inventor**
A person who thinks up and creates new things that haven't been made by anyone before.

**Laboratory**
A place used for scientific research and experiments.

**Lens**
A piece of clear material, such as glass, with a curved surface. Light bends as it passes through a lens.

**Life cycle**
The series of changes that a living thing goes through as it grows and ages.

**Mass**
The amount of matter that an object or particle contains.

**Matter**
Any material or substance that takes up space.

**Metamorphosis**
The change in form that some living things, such as caterpillars, go through as they grow and become adults.

**Microorganism**
An organism so small that it can only be seen under a microscope.

**Microscope**
A piece of equipment that makes very small things look bigger, so that scientists can see and study them.

**Molecule**
The smallest possible bit of a substance, such as water or air, that still has all of the properties of that substance. A molecule is made of atoms held together by chemical bonds.

**Nanotechnology**
An area of science and engineering focused on understanding and using materials on a teeny-tiny scale. It involves making changes on the level of atoms and molecules, which can't be seen without a powerful microscope.

### Nobel Prize
One of the international prizes given each year for amazing achievements in science and other areas.

### Nuclear fission
This happens when the nucleus (central part) of an atom is split into smaller pieces. It releases lots of energy.

### Observations
Scientists' observations are what they notice and/or measure in their work, sometimes using special scientific equipment.

### Observatory
A place, usually with a telescope, for observing things in space.

### Patent
A document that is given to the inventor of a new design to prove that they were the first person to come up with that idea.

### Perfume
A pleasant-smelling liquid made from the oils taken from flowers, spices and other ingredients.

### Periodic table
A table that divides the chemical elements into groups. Each element has a number that comes from the number of protons (one type of particle) in each atom of the element.

### PhD
An advanced university degree that involves years of study and completing a big research project. It can usually only be done after someone has already finished at least one degree.

### Physics
The study of matter, energy and the way those two things interact.

### Polio
A very painful viral disease that can cause paralysis. In 1952, a vaccine was made and polio is now very rare.

### Professor
A special kind of teacher at a university or college, who typically also does their own research.

### Recycling
When something old is turned into something new. It is a great way to reduce the amount of rubbish that we throw away.

### Reproduce
When one or more living things creates a new living thing, such as when humans have children.

### Researcher
A person who studies a subject very closely and looks for new information about it.

### Scalp
The skin on the top and back of your head.

### Scholarship
An award of money that students can apply for to help pay for their education.

### Species
A type of living thing, such as a dog or a tiger. Members of the same species are the same in important ways.

### Subatomic particles
A particle is a very small piece of matter. Subatomic particles are smaller than atoms, and can be found inside atoms that are broken open. There are many kinds of subatomic particles.

### Sunday school
A class held on Sundays, usually at a religious building such as a church. It teaches children about religion and sometimes other subjects.

### Tablet
A flat piece of stone or wood used for writing on.

### Tuberculosis
A serious disease that usually affects people's lungs and breathing. It is passed from person to person through the air.

**United Nations**
An organisation of many countries from around the world. It was created in 1945, and aims to solve global problems and make the world a safer place.

**Uranium**
A radioactive chemical element.

**Vapour**
Tiny particles of a solid or liquid spread through air, such as in smoke or clouds.

**Volcanic eruption**
When hot liquid and solid rock, and hot gases, burst out of a volcano. A volcano is an opening in the Earth's surface.

**Widow**
A woman whose partner – for example, her husband or wife – has died.

*Can you find francium in this list of chemical elements? Turn to pages 28–29 to learn more about its discovery!*
*Can you spot meitnerium? It's named after Lise Meitner. Turn to pages 24–25 to learn more about her.*

| # | Sym | Name | # | Sym | Name | # | Sym | Name | # | Sym | Name |
|---|---|---|---|---|---|---|---|---|---|---|---|
| 1 | H | Hydrogen | 31 | Ga | Gallium | 61 | Pm | Promethium | 91 | Pa | Protactinium |
| 2 | He | Helium | 32 | Ge | Germanium | 62 | Sm | Samarium | 92 | U | Uranium |
| 3 | Li | Lithium | 33 | As | Arsenic | 63 | Eu | Europium | 93 | Np | Neptunium |
| 4 | Be | Beryllium | 34 | Se | Selenium | 64 | Gd | Gadolinium | 94 | Pu | Plutonium |
| 5 | B | Boron | 35 | Br | Bromine | 65 | Tb | Terbium | 95 | Am | Americium |
| 6 | C | Carbon | 36 | Kr | Krypton | 66 | Dy | Dysprosium | 96 | Cm | Curium |
| 7 | N | Nitrogen | 37 | Rb | Rubidium | 67 | Ho | Holmium | 97 | Bk | Berkelium |
| 8 | O | Oxygen | 38 | Sr | Strontium | 68 | Er | Erbium | 98 | Cf | Californium |
| 9 | F | Fluorine | 39 | Y | Yttrium | 69 | Tm | Thulium | 99 | Es | Einsteinium |
| 10 | Ne | Neon | 40 | Zr | Zirconium | 70 | Yb | Ytterbium | 100 | Fm | Fermium |
| 11 | Na | Sodium | 41 | Nb | Niobium | 71 | Lu | Lutetium | 101 | Md | Mendelevium |
| 12 | Mg | Magnesium | 42 | Mo | Molybdenum | 72 | Hf | Hafnium | 102 | No | Nobelium |
| 13 | Al | Aluminium | 43 | Tc | Technetium | 73 | Ta | Tantalum | 103 | Lr | Lawrencium |
| 14 | Si | Silicon | 44 | Ru | Ruthenium | 74 | W | Tungsten | 104 | Rf | Rutherfordium |
| 15 | P | Phosphorus | 45 | Rh | Rhodium | 75 | Re | Rhenium | 105 | Db | Dubnium |
| 16 | S | Sulfur | 46 | Pd | Palladium | 76 | Os | Osmium | 106 | Sg | Seaborgium |
| 17 | Cl | Chlorine | 47 | Ag | Silver | 77 | Ir | Iridium | 107 | Bh | Bohrium |
| 18 | Ar | Argon | 48 | Cd | Cadmium | 78 | Pt | Platinum | 108 | Hs | Hassium |
| 19 | K | Potassium | 49 | In | Indium | 79 | Au | Gold | 109 | Mt | Meitnerium |
| 20 | Ca | Calcium | 50 | Sn | Tin | 80 | Hg | Mercury | 110 | Ds | Darmstadtium |
| 21 | Sc | Scandium | 51 | Sb | Antimony | 81 | Tl | Thallium | 111 | Rg | Roentgenium |
| 22 | Ti | Titanium | 52 | Te | Tellurium | 82 | Pb | Lead | 112 | Cn | Copernicium |
| 23 | V | Vanadium | 53 | I | Iodine | 83 | Bi | Bismuth | 113 | Nh | Nihonium |
| 24 | Cr | Chromium | 54 | Xe | Xenon | 84 | Po | Polonium | 114 | Fl | Flerovium |
| 25 | Mn | Manganese | 55 | Cs | Cesium | 85 | At | Astatine | 115 | Mc | Moscovium |
| 26 | Fe | Iron | 56 | Ba | Barium | 86 | Rn | Radon | 116 | Lv | Livermorium |
| 27 | Co | Cobalt | 57 | La | Lanthanum | 87 | Fr | Francium | 117 | Ts | Tennessine |
| 28 | Ni | Nickel | 58 | Ce | Cerium | 88 | Ra | Radium | 118 | Og | Oganesson |
| 29 | Cu | Copper | 59 | Pr | Praseodymium | 89 | Ac | Actinium | | | |
| 30 | Zn | Zinc | 60 | Nd | Neodymium | 90 | Th | Thorium | | | |

*Turn to pages 60–61 to design your own periodic table!*